C000272343

By the same author:

The Snowing Globe (Morten: Peterloo Poets, 1972)
Prehistories (Oxford University Press, 1975)
The Hinterland (OUP, 1977)
Summer Palaces (OUP, 1980)
Winter Quarters (OUP, 1983)
Out Late (OUP, 1986)
The Air Show (OUP, 1988)
Watching the Perseids (OUP, 1990)
Selected Poems (OUP, 1990)

THE ARK

Peter Scupham

Oxford New York
OXFORD UNIVERSITY PRESS
1994

Oxford University Press, Walton Street, Oxford OX2 6DP

Oxford New York Toronto
Delhi Bombay Calcutta Madras Karachi
Kuala Lumpur Singapore Hong Kong Tokyo
Nairobi Dar es Salaam Cape Town
Melbourne Auckland Madrid

and associated companies in
Berlin Ibadan

Oxford is a trade mark of Oxford University Press

First published in Oxford Poets
as an Oxford University Press paperback 1994

British Library Cataloguing in Publication Data
Data available

Library of Congress Cataloging in Publication Data
Scupham, Peter, 1933–
The ark/Peter Scupham.
p. cm.—(Oxford poets)
I. Title II. Series.
PR6069.C9A88 1994 821'.914—dc20 93-37268
ISBN 0-19-282337-X

1 3 5 7 9 10 8 6 4 2

Typeset by J&L Composition Ltd, Filey, North Yorkshire
Printed in Hong Kong

Acknowledgements

Some of these poems first appeared in *Birdsuit, Blue Nose Poetry, Festschrifts* for Edward Lowbury and Roy Lewis, *La Fontana, Poetry Book Society Supplement 1992*, *Poetry Review*, and *Rialto*.

Contents

The Ark

The boy on the floor is putting his ark to bed
To the sound of a sea which is blood, and sings at his ear
Of how it could ruck the carpet-fringes to weed
But leave those ochres and buffs unwatered, bare.
The blue of the ark is as blue as all he could wish for.

I watch the way he is watching his bridge of fingers
Walk hand-in-hand in the air with whatever comes.
A hold of promises, the ark grows stronger,
Hoarding its bitten features and spindled limbs.
It is hugger-mugger in there, a must of darkness,

And the rain is cats and dogs in a green outside
As the clouds and curtains crease into sudden skin
With sweats, curls, aprons, kisses: the moil and crowd
Of shadows aching for home or a clasp of sun,
A chirping of birds, a puzzle of half-built faces.

The kindly Ones and the Furies calm the air
To the faded browns of an always afternoon.
The mahogany table swells its feet, each claw
Anchoring sand. But now the rip-tides run
Till his fingers are only the flickering ghosts of fingers

Aching for rainbows, caught on the shoals of light
Swollen from uncleaned silver and blurs of glass,
The ark of his skull trepanned on the restless freight
That he must steer to its safe and certain loss
Through pulsing water, the long salt silences.

Less than the Truth

The rough dark goes solid. Trees come first.
Elm, oak, beech? They grow to the dream's height.
Fields tilt their hectares into camouflage.
November, teatime: cold quick in the nostrils,
Frost lacing foreground mud between the goalposts.

It would be good to be certain of this, to tell you
How a woman walks a dog on a slack lead.
That will not do, any more than I can show
The freckled captain with trailing laces.
That seat is real enough in its fluted irons:
Le vieux parc, invitations of ghost-life.

No. One must be content with less.
Nameless trees, fields, white posts
Gibbeting the air, mist coming down or rising.
I am sure of sodden leaves, but that is different.
They must be there, but they are not the truth,
Though, since they must be there, perhaps they hide it,
And why this feeling of panic, the encumbered soul?

Somewhere, trackers: the wet, dilated nostrils
Full of a coat-smell, hunting for half-dressed flesh,
A bandit, bent over the purr of his detector,
Out for the buried torque, a girl whose hazel-twig
Can leap nonsensically over the life of water.
Things here are so gone, unthreaded beads or needles,
Perhaps they should look by the trees? Look, not hope,
For the press of a split hoof, a clean feather.

Accident

1

Dear reader, I'm writing this to tell you
How a small freckled girl walked by the verge
With a white dog and her grandfather. Can we take this further?
You are dear to me, my lucky, astonishing friend.
We make rapprochement on the lip of sleep,
And though I shut my eyes, you keep yours open.
We do not read much at night in the blowing libraries,
But now, I like to think, her page lies open,
Slipping its catch between your beings and doings,
Tempting you to a view through its white window.
Here, then, is the girl, closed on her own meaning
As she walks from a glimpse to an image. Now,
She should be eating her dinner. 'Reader' is easy.
If I give you another line, will she be a sonnet?

2

But have you taken this in? There could be a problem.
In this line, perhaps one of cadence, that amour de voyage:
A girl and a dog, and her grandfather; yesterday's roadside
Blurred by the tune of a late-Victorian metric.
And how should a line like this hold your span of attention?
How was she dressed? Shall I tell you? How would you dress
her?
Oh, folderols, bright flows of things. Blue jeans?
Look, you know when you're in the front passenger-seat
And the driver brakes without warning, almost too late,
You swing, weightless and out of your assumptions
With a 'What the hell are you doing?' Well, glance in
The mirror. What's there? A clumsy line break,
A repeated 'white' in the first section?
What's happened?

3

Let's start again. Were you given this book? Did you buy it?
As yet I can't see its cover, or where it's lying,
Least of all, where I shall be when you read it.
What hubris, trusting us to have this future
In which to make a special sense of nonsense.
Let's get back to the business in my hand
And strip away those qualifiers: white, small, freckled.
Combined, they call to me with a tender disgust
As if I'm talking of a mouse in a laboratory.
And was the man her grandfather? Please forget them.
But though your limbo file is full of adjectives,
Remember the man who kept his treasure safe
By one condition. Whoever dug it up
Must not think of a white badger while he dug for it.

4

I'm telling you the truth. Because we *share*,
(Each is the Red King in the other's dream),
And each of us has space and time, can spare
A little of ourselves upon this theme,
It is important that we learn to trust,
If not each other, girl and road and man.
Do not believe the useful rhyme of 'dust',
Believe that I can make a sonnet scan.
A neat container: what then is contained?
A stopped frame from a film that's running free,
A coarse and grainy stuff that's not explained
By titling it in bold **'Life'**, **'Memory'**,
'Gone with the Wind', or subbing out the text
With flickering cartouches of 'What comes next'.

5

For what comes next is something we can make
Only in easy half-light and the half-rhymes:
Bright, broken things, the seductive lilt of the dark
Patting about with words at the old names—
August, with a fly-blown ache of sky,
Smelling of war and slow holiday,
With its Panama hat and a dolly-twist of straw,
Rough pasture, and a half-blind terrier,
A gravelled voice, pacing words slowly
And throat-stopped now for half a century,
A brass ferrule poked in a molehill,
The sag of wires from pole to pole
Black with birds and silver with sound,
The held hand lost in the holding hand.

6

How can I count her losses as they pass?
I cannot count my own. Each loss must be
Particular and endless. Does the dry grass,
At summer's end, whisper 'Remember me'.
There is a blur of dry and green, a car
Spacing itself to where the day begins—
They're something still, but were not what they are,
Such particles have glimmered from their skins,
Each photon stuck and hinged inside a head,
Probing the space–time of the zodiac
Or lying down, pretending to be dead,
Beyond the spooky half-life of its track—
Reader, what do you see? Let's play 'I–Spy',
Scrambling the skylights of a single eye.

7

Let's take a sodden castle, in a rain
Which mizzles blue and unremembered hills.
Stand on the magic carpet. Count each pane
Of puzzled glass. The rocky landscape fills
That crazy paving with the ghosts of trees
Which meet the ghosts of chiffoniers and clocks
And offer up their crystal truancies
To light which greets the window, and unlocks
A living space: one bright, and real, and *so*,
Where loll-tongue dog and man and chattering girl
Can meet in fiction's radio-active glow,
Conspire to make our head and heartaches whirl.
Outside and in don't match. A missing room
Hides dust and lanthorns, mémoires d'outre-tombe.

8

Is it Lot 93, or 92?
The sharp-nosed auctioneer, two lots a minute.
'What am I bid?' No provenance, no signature.
A rural scene. Amateur, but not a print.
The gangling man in leathers, the holiday-makers
In from the rain, and Mrs Moore, dealer in bric-à-brac,
Are silent. Everything must belong somewhere,
Be at home in a finite series of numbers.
It's a rotten likeness of something, but still a likeness,
And will grow lovely if you grow to love it:
An old man blurred enough to make an ancestor,
The dog and child whose trust, fidelity,
Is all you feel, and all you feel you've lost,
The frame waving goodbye as they walk towards you.

9

There is merriment and puzzle in Cookham Churchyard
At the moment of resurrection, when all discover
The relief of realizing they need no adjectives,
Or need them only in the Pickwickian sense,
As nuns own maiden-names, or rosaries.
No one says 'What a funny hat you're wearing!'
'That walking-stick was Jim's.' There are no brand-names
For brands plucked from the burning, the saints and martyrs
That we have dignified to be our votive images.
And if the Panama hat, the aertex shirt belong,
They must belong as prayers and flames belong,
Clothing the lost majesty of Proper Names,
Which, at their most proper, become as common
As man and girl, as dog, as road.

10

The underwriters tell me I should tell you
About this accident, this impact. I drove within my limits.
There was no tyre-squealing; there were no skid-marks.
All that humming cage remained responsive;
Just a catch in the engine's throat, a change in the fine-tuning.
I pulled the car in carefully to the verge
And remembered the hazard-warning indicators.
A bit too late. I realized I'd gone clean through them,
But the sun picked them up in its own good time,
Walking no more, and no less carefully.
I think the girl might have been carrying flowers.
What do you think, dear reader? You're the only one
Who has some cause to make a claim against me.
Is our cover comprehensive? Shall we call it knock for knock?

What do you Believe?

The voice of the dying man spoke from the marriage-bed,
His head lit by the lamp closed on its dark wick.
He spoke, dying, between the rock and the dead light
From a hard place, where the blackness had no seam,
No rivulet along which there was movement,
Or space in which one seed of time could grow,
As if that lamp, in its going out, could speak
From sand, countless; from disinheritance,
Sending the stir, the shock and bloom of things,
To beg its way from flesh to memory,
From memory to image; then to nothing more
Than the one question: 'What do you believe?'

Someone there at the bed's foot, spoke to the sheets,
To the pale eyes wide open on the dark
And all that sand and disinheritance
Of woods and waters, closed on their glittering,
As if there were words to answer those four words,
The lamp could be trimmed to burn with its small flame
And make more luminous one serious thing;
Picked up a book, and read to those wide eyes
Words by a dead man, in which the inconsolable dead
Had looked for consolation. The wasted hand
Which could not move stirred for the lamp's key;
Set the globe leaping with a clouded light.

Spoon River

'It's Spoon River', he said, looking past things
At refractions of marble, swinging in chains of light.
'And if I could tell you, what could I tell you?'
Best to say nothing, as they were saying nothing;
Do nothing, but look at him looking:
A head worn into patience, a kind of godhead

Hunting for space between judgement and mercy.
We turned away from the heat of the graves
Where his mother and father lay, his brother,
In a dark dream of grass, brightening
In a late sun, in a time before time to go
From this clock-stopped time of gold and edges.

Was it playing God? Each to his own omniscience,
Which was denied. Did he call for resurrection,
His head a stoneyard where a lid thrown back
Set Billy Brumpton harnessing his mare again,
With wisps of hay and cooking-smells at corners,
The sound of old money arguing at market?

But Spoon River, with its hanging judges
And women taken in adultery?
He had nothing to say to this town in its water-lights,
Its blobbed flowers dying in handfuls.
'If I told you their stories, it would only be lies.
We shall leave all that for the Recording Angel.'

No locks on the gates. As I turned back with him
The light grew confident, then magisterial.
What could be said? To whom could we say it?
Their lives reached for his keeping; his unmoved eyes
Gathered their names as a child gathers in secrets,
To hold something in trust for the trusted silence.

The Pond

The sheep hangs on the surface of the pond,
A porridge of bobbled flesh; the trees are still,
As still as anything. We stand there, and look down.

I had day-dreamed this moment, showing them off
That clear dark eye; the fence of thorny stuff
And all the careful crookedness of paths

Leading to where I had only asked a place
To be itself, and keep my kind of secret:
Lit water, holding itself against the rim.

This was the place, refusing to come clean,
The sky, spreading its headache over the dip
Where zig-zag lights and weary filaments

Picked at the ragged edges of the leaves,
The natural day scorched by their dirty brilliance;
The secret, turning its insides out for us.

Long snows, these winters. The packed cold
Mulled by black boots into rot and grey,
Yellow animal piss in steaming holes,

And, overhead, the nursery rhymes go on:
A cloud like Mary's lamb, led to the slaughter,
Snow-goose feathers, plucked from something dying,

Which pass and settle, settle and pass again,
As I pick my way, eager, the lost leader,
Towards the ambush and the long way home.

Luigi's Garden

Night slides off the Umbrian hills
And fireflies come out in Luigi's garden:
Smoke-dark, high-lit, the shallow stream
Trips under the bridge to feed monlight;
The sky ripples, on the edge of lightning,
At the dull thud, twang of his music-system.
But the Mills Brothers, with fur-lined voices,
Try to fit them to the Glow-worm Song,
Dazing Linnaeus with close harmony,
These lampyrids, elaterids,
Not to be confused with Lampyris noctiluca,
That heavy girl crouched in long English grass
Giving the green light to her blacked-out mate
From an abdomen of uric acid crystals.
'Thou aeronautical boll-weevil,
Illuminate yon woods primaeval'—
Enough of nature-notes and country comets:
Each shiner is the other's metaphor,
And as for 'lighting tapers at the fiery glow-worm's eyes . . . !'

The *Dream* again? Yes, the performance,
As always, wrong in its open-air production.
This summer, every night's a first night,
And, perhaps, the backdrop crinkled paper,
The bridge a merely operatic construct
Strayed in from *Cavaliera Rusticana*.
Max Reinhardt strings his fairy-lights
Across the insubstantial gauze of evening,
And now that Micky Rooney has stopped kidding
We can put a few lost lines together,
As if the play were a charred manuscript
And all that can be read is 'Night and silence',
What night-rule here about this haunted grove?
Merely the wand of darkness tipped with stars,
A twist of proper names. Titania's power
Diminishes to the chime of Tinker Bell
As these faint pulsars navigate the dell
And we give audience, who know
The best in this kind are but shadows.

As all our lights are only shadows
Of the one 'Fiat lux' that it is necessary
To put our faith in. So, in this other country,
In the shade of an old mill which grinds nothing
But a handful of dreams and idle summers
Back to the chaff and grain of memory,
We stand and watch, without really watching,
These macula, indifferent particles,
Scraps of glitz, wisps whose will
Wanders free from our determination.
The music off, the house-lights down,
Out of the hush, their stammered off-and-on,
They fly as tributaries to our parables,
Remind us of our singled affirmations:
Smiles for strangers, reassuring hands,
The small courtesies of a meant 'Good morning',
Civilities of the sick-bed, gifts out of season
Making a dance where windless hills incline
To one small festival in Luigi's garden.

Fête Champêtre

How to welcome the advent of remarkable things?
The King in his double-dyed-in-the-wool wig
Steps out with a mobled Queen, ripe in full fig
Walking as if on water, arms akimbo—swans' wings,
  The hissing stifled; the beaks pointed, poised,
  The lake stage-struck by rain, and all amazed

At the tenor miming his love in a dark gondola;
His flab abeat with an impossible desire,
The Tannoy going Oyez, Oyez, mad as a Town Crier.
We sit it out on a wet night at the opera,
  Watch the sweet nothings, listen to whispering grass.
  The rain has brought things to a pretty pass

As each damp dryad twirls in a daisy-chain
Across ridiculous trees; a summer frost
Of limelight crisps the edges of all things lost,
Drowned and unchancy: plane across rainy plane
  Whose greens cannot be followed a thought further.
  What could, in this velvet hour and this weather?

You can see through the clung dresses down to the bone,
And through the bone to the marrow of the night,
Whose rich pith shines by sizzling sticks of light
Planted in mire, guttering, dying alone
  As each cold courtier must. Let tender rain
  Fall softly, softly on all our Castles in Spain

As the rockets kiss goodbye to cars in echelon
Hull-down in mist under each slow, red rose
Of stars lipping the dark with their oh-so O's
Until the last crackles of silver rain are gone
  Which lit the hampering hampers, the watered wine,
  Leaving us leaving, leaving things left again,

And nothing more extraordinary here but rain
Building its tarnished scaffolding as close air
Swells, adrift with tears, brooding on where
Those gay umbrella-segments bobbed over real pain
   Which knows its place, has given a long slip
   To this littered ground, which looks like clearing-up.

Cat on a Turkey Plate

It is a fine thing, and a sufficient thing,
To see a great cat of weight, some fifteen pounds,
Hoisted, in season, onto a turkey plate,
Quickset with holly, lord of his own grounds

Ancestral, with a flame to burn behind him:
A cat of amber, with full and electric fur,
His eyes open, but not with the wax greens
That light red candles, and to hear the great cat purr.

Such a calm breath, such comfortable stance,
And all around the small sighs of applause
At the Christmas Cat in his scallops of old china,
The needles working from the slow flex of his paws.

Homage to him, in the deep pile of his carpeting,
His bend wavy, though not sinister.
He waits for hands, and the paper-chains of love
To minister unto him, as he does minister.

And though the wicked knife and fork in the half-light
Can see things clearly, and can see them plain,
In his benevolence he dreams of making
Umbrellas for mice, to keep them from God's rain.

It is a fine thing, and a sufficient thing,
To see him at ease, large and admirable,
In a room powered by his own true carolling:
It is a grace before meat. It is seasonable.

Monet's Garden

The old man stepped out of his studio,
Ridiculous tin car, unfinishings,
Leaving some dignity of beards and ladies
To keep the past and its bucolic engines
For those to whom black is a favourite colour,
Who drain themselves back into photographs.

He asked if we had come to see some flowers,
He'd tidied up the kitchen, scorched it yellow,
Invited Hokusai and Hiroshige
To climb like cats on every wall and stairway,
And told his pictures how to turn themselves
Into the sizes that would fit our pockets.

His flowers rippled gently in the sunlight.
He sat in the green boat he called the garden
And rocked it softly. Crumbly at the edges,
The shadows whispered in a foreign language,
And all the flowers we picked and pressed in childhood
Were there for us to pick and press, in camera,

As white and bridal as our mothers' trousseaux,
As pink and blue as an Edwardian nursery.
The lily-heads came floating up at us
Out of that tumbler full of painting water
In which we used to dip our sable brushes
In long and hot imaginary summers.

He waited for us under hoops, down alleys,
In a pink house pinned up with open shutters.
What could we say to him, to Monsieur Monet,
Except we thought it pretty as a picture?
He thanked us, out of radiance and silver;
A ghost can make white light out of a rainbow.

The Web

A broken web is hung about with rain,
It could be perfect, but the perfect hour
Hangs in the gallery of God knows where.
Things pause, and stay a bit. Wet stone, white flower

Make no alas for that, but being so,
Catch at an eye that does not look for them,
And presidential in the morning light,
A quick-still spider, cloaked in diadem,

Threadbare, indifferent till spinning-time
To those lost finishings which looked like art,
On props enough for its own truth of things
Holds the tight epicentre of its part

Where lust and hunger turning into green
Become a garden: that menagerie
Of pick and unpick, hot and wet and wings,
As right as rain, as sun, as things could be

Which batten on the borders of backyard
Where nothing seems to fit; just lies about—
That mountain-bike with rusty handle-bars—
And nobody much in, and no one out,

Or someone's landscape cooled by balustrade,
With stencilled cows at munch among the rough
And lawns and lawns, parterres, and lawns and lawns,
A bowl of Chinese fidgets—all that stuff.

Clouds come, clouds go. They never look much wrong.
The fly gets framed, gets trussed. If there's a soul
The web is hung how something wants it hung
And would not be more perfect were it whole.

Half-Hours in the Green Lanes

Half-Hours in the Green Lanes, by J. E. Taylor:
'The first book that I ever bought with my own money.'
Birds, whose sharp cries relieve lanes of their loneliness,
The feathery tribes, the pleasing recollections—

'If you want my autobiography
Read Wordsworth, Traherne, Vaughan.'
As Donne says, 'There is not so poor a creature
But may be thy glasse to see God in . . . '

And, in his commonplace book,
This litany of conscience:
'Bird: A Penitence. (Et Ego)
A Blue Tit, A New Air Gun.'

John Woolman, who killed a robin with a stone
Then killed her brood, who had no dam to nourish them.
'The tender mercies of the wicked are cruel.'
Bewick's bullfinch, his little Matthew Martin.

Frank Kendon's catapult, the shrilling wren
Whose entrails lay shining in the hot sun;
De la Mare with David's sling, a sparrow for Goliath,
'And the hot, pallid, grinding shame of it.'

Charles Tennyson Turner: a swallow,
Its fallen head huddled in highway dust:
'I would not have thy airy spirit laid,
I seem to love the little ghost I made.'

Taylor: 'Their movements create an animation
Whose absence we can hardly realize.'
I remember his childhood's case of butterflies,
Dead freshets, rusted pins and crumbled thorax.

Annunciations

1

After siesta, the promenade, the passeggiata,
And Gabriele looking for Maria:
That easy body-language, shoulder-shrug,
Flexing of the parti-coloured wings,
Plumage stroked down by light
Which runs like water off the oiled feathers

Still warm from such a golden plummeting—
Heaven's courts into the quattrocento.
They're making notes: Piero della Francesca,
Spinello Aretino, Beato Angelico—
A whole host of them who paint like angels,
Can draw celestial circles freehand

And fix the pair into their double-curve,
Balance his pinions on her reticence.
And should she laugh, throw her pretty hands down,
Would they catch the rattle of his lift-off,
The hand slammed on the hand of a fellow-angel—
'She wouldn't listen'—the drifting primary?

2

Wind rose to its full height.
When I asked for a sign
The dead leaves hissed
In their cold plumage.

'Whose child is this
I was born to be?'
From a blue-rimmed saucer
The tea-leaves answered

In dreary twists:
'Of chance and happenstance.
Their long bickering,
Unfeatured faces.

They have entertained you
Without entertainment
And unawares.' Then the shiver
Ran down my shoulders

As if the want of wings
Could grow them;
There was a message
For me to deliver.

3

And these interludes? Angel-visits,
Few, considering how great the hosts;
Far between, as interstellar spaces.

He flew in from the sun, grass brightened.
There was clover, there was speedwell.
Hundreds of names lay scattered broadside

And yours was there, and yours, and yours.
There was neither promise nor recollection,
He stood there on the unbent flowering

And someone took a photograph.
Such a whirl of shadows; the cold
Chipping away at toes and fingers.

Here it is: a perfect tabula rasa.
Was it the happiness that wrote so whitely,
Was it the deep illusion's perfect zero?

4

I knew him straight away, my guardian angel,
I counted on him in my hour of need
To be my friend, my mentor, show me how
To make a blood-bond with our Saviour,
Redeem the promise of my white election.

I saw by inner light how bright his eyes,
And how like mine: that north sea chill.
We guessed our thoughts out by each other's whispers;
Spilled each other's tears upon the streets,
His vice-grip tightening my virtue.

The blood-beat thick in my anointed temple,
I felt his halo burning sulphur-yellow,
His keen wings a jostle of bright knives.
I plucked a pinion with a razor's edge
And set about to purify the city.

5

It was drifting about the woods all day
In desperation. The birds mobbed it
Through thickets of 'Halloo' and snapped branches.

Perhaps it came from fields of light so bright
Our half-light dazed it. Someone drew a bead on it
And watched it fall back into fallen nature.

Pulled out into a roughish crucifix
It hangs now in the gamekeeper's larder:
You might just think it touched by evening sunlight

But the gold is real enough; rubs off like pollen
On the plain skins of those accompanying thieves:
The magpie and the stoat, who swee there, drying.

And that so-human head. Does the wind turn it?
What maggot-life gives those lips a credence
As if they had a message: urgent, undeliverable?

6

She bent over him as he lay there,
Wasted to androgyne, the drip feeding
Its pure elixir into impurity.
Time on the clock-face ran
Past absent numerals.

Then sat, her worked fingers
A light cross borne on blue
Which opened onto sky and summer.
His heart paused at a familiar door,
Then, with quick patterings,

Crossed the green lino to the bed
And waited for her.
'Lullay, lullay, my little tiny child . . . '
There was a cold as snow and soft as feathers.
Smiles lit the rooms

Whose eight walls folded into four,
To the run of curtains.
The clock spelled nothing,
Spelled it clearly:
Only a face kissed by hands,

7

Azrael, whose wings are of a muffled sable
And undistinguished by the life of stars,
Is attentive at the last bedside.
History has not been kind to him,

Who has cloned himself so often,
Wearily uncloaked his case of opiates.
A Phoenix, whose eyes are benignant tumours,
He rises now for the last time

To make an end of suffering;
To lead those cooling and transfigured hands
Through curtains of black velvet
Onto the morning pavement.

With a wave of his hand,
He stares into the forbidden light,
Waits for the second trump of the Archangel,
The uneased accomplishment

Of his mortality, the wings
Blown leaf-dry, moth-dry,
The dust of his passage
Sun motes dancing over New Jerusalem.

~

Foliate Head

Deep in, the unintelligence of things:
Coarse lines, underscored,
Knit rust: no hold, vantage.
But the painter perches en plein air,
Smooths into light his gratified desires
Of wandering mothers, bonneted children,
Whose flesh is drenched in rose-shade.

Air sweetens his glancing brush:
Twist of sabled sunlight
Gold fan in a blue-veined hand,
Poised as the light poised on a lace-wing.
Today the woods are nothing more
Than the cool end of a spectrum
Where trespassers must be prosecuted

By what's at home in ambuscade:
Branches at their trammelled coronation
Of a face simple as a wicker skep,
Natural as the whip and thong of bramble.
Out of those intricacies, features
Fly into place as tesserae fly to a wall,
Patching fired fragments into sainthood.

This is not sainthood. From the selva oscura
He watches the artist pack his keep-nets
Loose with smiles, fleece-grass, featherings;
Rooted in metamorphosis,
Would take those hands, work the nails to buds,
A transplant made, a sharp graft taken,
His heart of darkness centring the light.

Stone Head

The tower so clearly made of light,
The loneliness of light, from lonely money
Sighed away on a late medieval death-day.
And the pretty names: mouchettes, sound-holes,
Double-stepped battlements with pinnacles—
One of God's fairy things

To sway with clouds, but pinned
Into the illusion of a high motion, lift-off,
Setting four devils at their platform angles
Reaching for Heaven, crop-winged
For fear of flying: saturnine,
Staring themselves out of countenance.

Till slam. After so much star and cloud
A thunderbolt of wind cracked off the corner
And flung this one poor clone of Lucifer
Headlong. Picked from couch-grass,
His down-to-earth bone-glare
Domesticated in a red-brick alcove,

He has put off the smell of brimstone:
One of the Foliots, Lares, Genii,
'That kinde of Devils conversing in the earth',
As James puts it in his *Demonologie*.
Lop-eared, crook-mouthed, cheek-chapped,
Kissing-cousin to Meg Merilees.

The Longaevi, the fallen angels:
One down, and three to go,
To suffer jolt and jar or slow erosion.
Look in the grass; find his bolted feet
Still clawing for a toehold
On a flawed block of consecrated stone.

Nacht und Nebel

for George Szirtes

1

It seems, at first, as if the eyes of the dead
Are sleepy with sadness at being first to go.
What was wanted, not wanted, doesn't matter much.
The muscles of life become involuntary;
Leave the eyes as open as that window
Traipsed by cloud. It has been met now,
Whatever *it* is: mist, night or radiance,
Those words we choose to make a special nothing
Of nothing special. Flesh pared to childhood,
A parting gift from things unparcelling?
That I didn't see. Childhood was in the tears
Cried lately, late. As night comes tidying up
A face begins its pilgrimage of mist,
The loosening features bandaged into memory

And love: that word with its intolerant echo,
Loud as the gibberish racing off sprung-concrete
As the coach leans into the tunnel's curve
Stripping the hunted rails clean of silver
And grimy windows fill with passengers.
The family-faces drift in fuzzy light;
Turn towards you as you turn towards them.
There is a weakness in such inflammation,
Arms loose, heart shaking in its cage of bone.
To love the dead? That is cloud and nocturne,
A yes that is the negative of no, a bouncing bomb
To breach the dam set at the valley's head:
Landscape fills to an eye, headlamps under water
Racing for out, the drowned rising.

2

Clocks are the unblinking eyes of time,
Cyclopean: eternity must blind them.
Hours are numbered carefully, forgotten,
But, in time, every clock will cover
Its white face with its hands.

Here is George's painted wooden one:
Mors, Virgo; Death and the Maiden.
He takes her liltingly from behind.
Soon, her bones will penetrate her skin;
Now, her hair flames to a furious sky.

Between the starry floor, the watery shore,
Reds and ochres talk of Bosch and Breughel.
Saturn, roughest of luck, tosses his head back,
Mumbles a child's head with hot tonguing,
Another kicks lightly from his idling hand.

If there were trees here, as in Tacitus,
A centurion and his rape of soldiers,
Numbed by the silence of the Schwarzwald,
Could watch dead legions hanging on live branches,
Their arms grounded, the earth souring.

In such an accoutred landscape,
Taylor, the Water-Poet, saw the hangman
'Burst his chin and jaws to mammockes,
Then he tooke the broken, mangled corps
And spreade it on the wheele'.

Let the meal of suffering be garnished,
The pangs of wit soften a deathbed.
Pastiche, rococo, petal that dark motto:
'Timor Mortis Conturbat Me.'
The clock's brazen heart beats faster, faster . . .

3

Iritis: an inflammation of the iris.
Take a single eye, swollen, under pressure,
Meat-red. In the absence of atropine,
Bella donna, the lady of alleviations
To dilate, fix and school her eager pupil,
The eye rages for a draught of night,
Hungry for mourning, sables, cold velvet,
To be buried alive, out of sight and mind.
The visions are tolerable; they are of nothing.
Sleep? Your corded eye, a giant bladder
Will not admit you. A pure and endless darkness
Is the one unsure promise you want kept.
Let there not be light: one splinter of white glass
To break the spell of your blind date with pain.

Old Street for Moorfields: all those patient backs
And turbaned faces. On the optician's card
The crownèd A is quite recoverable
Over small anagrams of Amor Vincit Omnia.
It is good to bring things into focus
And grow accustomed to mere double-vision,
To find Blake's old man crouching in a thistle,
His guinea disc a host of dancing angels.
There is the natural pick and mix of metaphor:
A festival of clouds chants *'God si love'*.
While Dr Aziz flashes his instrument,
Looking for secrets in Miss Quested's eyes.
An ambulance breaks the lights in mounting panic;
Newsprint feeds the famine of the gutters.

4

Dying is so much, so many of it all,
Each pair of eyes that opens, looks and dims,
Making its lie of sense, its lie of nonsense;
Guiding a pair of hands to tug the world
To the same difference, a tongue to cry 'I love you'.
It is as much, as many as being born;
The one perfection is that equipoise

I met Lazarus; his name was on the Menin Gate.
The Last Post had been blown for him so often
That all his tales lacked credence. He had come back
From that obscurity which is the Kingdom of Error,
And all he could tell me was that he was dead.
Now, at the end of April, not far from Easter,
Something has come again to make you shiver,

To set your teeth against anticipation.
It would be nice if things went swimmingly in heaven
For kings in pyramids, our common clay,
But who could bear an eternity of memories?
Someone's up there, schoolmasterly, round-spectacled,
Telling the Schützstaffel how to mark all papers:
'Nacht und Nebel—night and fog.'

5

Pencil and film: our two recording angels.
In a quiet hall high in the Pompidou Centre
Ghosts from the death-camps lie on ghost-paper,
Expose themselves to gentle, special light;
But the old photographer is a spool of bones,
His camera's lung is punctured, its eye clouded,
Shuttered down on a host of negatives.
In the houses of the dead reversals of light
Char flesh, burn off the frocks and jackets,
Chalk out a court of shadow on grass
Cut to a level killing-field. What remains?
These tricks of camouflage, arguments of ash,
A child's case packed in an Auschwitz warehouse:
'Jungkind, geboren 1933'. My year.

And how about some bricks for that child to play with?
In my grammar of architecture, alphabet,
Things were spelled out in black and white,
The white lie making its grey bridge between them.
Allow his bricks their little absolutes,
Coded as we code devils, rats and snowdrops.
Black, grey are safe: fear the presence of white
Which is a little too much like the radiance
Of lightning, bomb-burst, bone and sacrifice.
The child sings, building its fragile city
Out of the neutral tones of a dead war-film;
When things are put to sack, the tower-blocks fallen,
There are always crying women, shawled in black,
A white arm skewed from the pit, grey rubble.

~

First Things, Last Things

And it was first, after the mess and straining,
Runs of hot water, new-bread smells of linen,
Ripples of light, and I think there would be flowers,
To bring, from throat and loosened tongue
The urgent vocables of love and fear,

Which might be only the ungiven names
For what was warm, cold. The night, the night,
A face pressed against the press of milk,
And, under a smile stretched from her smiling,
The suck and sob of things . . .

In between, the long brocade of lies,
Truths, half-truths, webs of such distinction
That every face he knew could grow to fit them,
Though there were flowers with a myriad names
Dying easily on dying grasses.

And it was last, and the retreat of words
Drained from the bed, as Matthew Arnold's sea
Drained, still drains over the smoothened shingle.
Again, flowers: the new-bread smells of linen,
The urgent vocables of love and fear

When hands that never touched were quick to touch,
Touched to the quick. There were no names for things,
Because, in that communion of shades
Which sang in paradox with angel-voices
The dark was entered easily as the light.

Your Troubled Chambers

It's there, in a flash, a fist crouched on its anger,
A dark blush stumbling the cheek of a lover
Caught again in the toils of his recognition.
You were there. What happened? Time may tell.
He tells lies easily: white lies at a garden door,
Darker lies which breed like rats in a coven,

Their blind eyes nuzzling for truth in a crowded well.
Go to the weird sob-sisters. They might help you,
Chancing their witchy crystals on kitchen tables,
Teasing out lifelines, measuring you for a shroud,
Snipping strength from your hair, gabbling in dreams—
Blank verse which stings like soap in a child's eyes?

What works, works in the dark. Things wait their event.
The child in the netted pram smiles at her killer;
A nuclear trigger clicks on a bare-ribbed plain,
Its giant camera snapping the troopers' bones
And fingerless halt babies wait to be born,
Crying softly from the other side of nowhere.

Back in this community of the faithless
A schizophrene hoods his eyes on the pain of light;
A surgeon bleeds through his gloves and instruments,
Signing a death on the horoscope by your bed
And something sings 'Have you no ghost-room for me,
To let me second-chance your troubled chambers,

No room, perhaps, for a drifting chromosome,
A patch of drying blood and spun-glass hair
Once pasted on a wind-swept yard of tarmac,
An uncle whose sweet caresses touched the quick,
Or such slight ectoplasm as a wagging beard,
The wrong light on the dust in the right room?'

Front Parlour

No doubt, the room could hear what was going on,
Or, rather, sounds could pretend themselves at home
Without disturbing a leaf in the jardinière.
There were lilies there, which neither toiled nor spun,
And roses worked to the stems in silk, for time
Was a bunch of immortelles; had enough to spare

Of itself to web, delay those messengers
Whose messages, unforgotten, remained unread.
Those *Four Just Men* and *Girls of the Limberlost*,
Safe behind glass, looked strange as any strangers;
A cast-iron Norman fireplace burned the dead
With invisible fire as hard and pure as frost.

And everything there knew it must be for best
To be still multiples of their quiet selves,
Consuming air, as the votive candles do
In a cool temenos where God is lost
In his own dark shawls of wonder, the salves
Of a dying silence let something through

Which was not the sound of children on ricketing stairs,
But the rustle of painted leaves on a balustrade
Of painted cloth, where the Passchendaele soldier stood
Waiting for Charon to ferry him down the years,
The spick shroud crimped for the mantelpiece parade
And all kept safe under the veil, the hood,

While the net-curtains eased shadows about the light
As a green screen parts, unparts, over a glade
Littered with tiny icons of skull and bone
Where the dry moss pads about on its mild grey feet
And the uninvited bomber scrawls overhead;
And all the desires are here, and are clean gone.

White Lies

Is it the saying, for something must be said
As a livre de présence, an enunciation?
Rather, for her, an annunciation: the words
Tucked loose in the dove's ring, the good tidings
To let the tired leaves know how green they are,
Her calls consoling as the lightest rain
And that good-earth smell of transfigured dust.

What does she see as she follows the garden-path?
Flowers at a ticker-tape welcome, the whole panjandrum
Alight with sun and unspooled messages.
The cat, that good thief, told he's a good cat,
Cries over spilt milk in pure delight,
Trembling a dark purr. And then the pleasure
Of greeting them all again, those terrible visitors

Who have borrowed the habits of annealing shade,
Their hopes as dull and thick as furniture.
How well they look, and how each colour suits them.
She must offer them a summer of Christmas cards,
A winter of wings beating the butterfly bush.
What more can she do, to make more beautiful
Their faces lined with plain, unvarnished truths?

But ask them to come again to this house of wishes
And read their welcome lurking in the tea-leaves,
Their lucky fortunes in her crumble-cookies,
Or, lost on the doorstep, look up to the bedroom
Where the suspicious curtains draw their secrets,
And hear, as she lies there crying on her pillow,
'I'm afraid she's just gone out': an occluded star.

The Clock and the Orange

In the living-room, the dead room, full of anger,
The two of them stand, each to each stranger
Than when they first looked, each at each, closely.
Wondering. They have filled it quietly, furiously
With choice and argument: rug, chair and book,
The wedding-present of a careful clock.

He swings it head-high, lets time drop,
Cower on the floor in its time-slip.
Will he throw the clock which is poised there
With its delicate wheels which have come so far,
Which has kept something in some sort of order,
As if a clock could draw the line at murder?

She stands there facing the light, the light facing
Up to their faces, squaring up to the amazing
Thing which can't happen, is about to happen.
She grips a foolish orange, her eyes open
As a hung mouth. You can just see the light
Which is dead-still, and can't get out,

Glazing the clock, so brown and circular,
Its reasonable tick trimming the wild air;
The passionate fruit with its tawny rind
Locking its love-juice round and round:
All wet with tears, a closed bubble,
A marriage stupefied to lies and parable.

Light Frozen on the Oaks

Light frozen on the oaks: a long time without wind,
And nothing much to mark this morning, except time
Hung in gold letters on the spines of books,
Dust-motes claiming title, a long-case clock
Stopped at the hour past the eleventh hour,
Its hands surrendered to some noon or midnight,
Struck among roses and unflurried birds.

The sky, in all its graduals and ordinaries,
Promises no more than a midwinter pallor,
A reach of spirit to a shelf of silence
Where memory has arranged its broken secrets.
Even the dead must hold their breath today,
Their texts a commonplace, their strength of purpose
Caught in an act which breeds no consequence:

Johnson, the axe of some deep sentencing
Swung to no fell; the table-talk of Coleridge
Stopped as a fountain in the heart's wasteland.
Today is a garden of dried flowers, their scent
Far colder than that reach of pink and blue
Unrustled in its china cabinet.
Are you eager to watch grass grow, light move

In answer? But the shadowing of the sun,
Delays its fingers on the skin of things,
Sleeps in the cracks between was-once and will-be;
And the clock, chained to conspiracies of wheels,
Is birds and flowers programmed without memory,
Without hope; as the wind, when it comes again,
Will deliver its empty mouth up to our sighs.

The Provisional Thing

And again, the provisional thing
As everything is, provisional—
In default of that Arcadian wind
Which could cut flower or skin without such sneaping
As this, from the East,

Which works over the ground
Without promise, without adjective,
And has no bargains to make with us.
There might be time, if time is of the essence,
To fill the wind in,

Provide it with a season,
A change of direction,
The blobs and runnels of a loaded brush
Blurring the contours of a simple landscape
In a simple book.

For now, it will have to do.
It must know its way about
After four hundred years of hole-and-corner work,
Fossicking at brick, falling down, sobbing
With the hurt of it,

Borrowing the pain
Of all things loose, discarnate,
That have borrowed our incomprehension,
That sing for the space, provisional,
The house is in default of.

A Habitat

1 All Roads Lead to it

There is always a guide, a golden bough of keys
Loose change in his hand. Leaning on the car,
He looks at netted thatch, close brickwork,
The plywood lids nailed across the eyes.
In a cold foyer Virgil waits for Dante,
Looks at his watch. 'Shall we go in?'

And what will you learn, when you find your survey
Was 'conducted in virtual darkness'.
There are, as usual, major areas of instability,
Widespread infestation, structural distress;
The seven standard requirements have not been achieved.
'I am not fearful of major collapse in the short term.'

Though, perhaps, one should be, thinking it over.
Who bewildered those floors to sand,
Tumbled the tumbled brickwork? Who attended
To the long marriage of queen strut, king post,
Set swimming in the porch's pediment
A merman and a mermaid of white stucco

Who swing their lovely, rough-cut tails apart
And stare with chapped eyes at the east?
Who sing, as lorelei from their high storeys,
Over lost time, over all loss and time:
'We have this brilliant magic: to command,
Cheat dirty seas, and raise the drowned we drown.'

2 *Out of Season Harvest*

A cock-eyed house, beset by open fields
And too much wind. Each evening,
Light, undressing, makes itself a ghost,
Slips through chipped hands, looks dull,
Pores over unfaced brick and threads of lime.
In these worked-out seams, poor stuff:
Nettle-rash, and casts of brown iron,
Small saws of grass to cut, and come again.

Spring, in all its brutal innocence,
Webs it about with marker, stake and claim.
Is there a garden for those teeth to bare,
Disclosed, but secret, all its petal-shells
Rehearsing rumours from a summer sea,
A child, blood-sucking from a slashing blade,
Blown into rags of travelling light and dark
By the same wind, though swollen with lost voices?

There is forgiveness here; not much forgotten.
A simple, careless weight of young and green
Saps ill-seasoned wood on its chewed hinges.
Across the empty spaces you can hear it,
The old lie of the land: 'I do my level best',
And feel the farm hands, calloused, broken-in,
Unclog the shreds of yellow binder-twine,
Leave used-up life to dry out in the sun

For you to glean; an out-of-season harvest.
Then, bent-backed, feel the house hang loose,
Flag in the wind, flock of substantive shadows
Punished by echo and the ghost of echo,
The need for ardour and an arduous love.
What can you do but take it to your arms,
Undress its walls back to the mother-naked:
Each worn to ravellings, locked face-to-face.

3 *Brick on Brick*

They have put brick on brick,
Those headers and stretchers,
Hands mapped, wrinkled
With white scar-tissue.

Houses, common as dirt,
As moonlight, sunlight,
Slide their counters
From lean horizons.

And much sweet-talk
Of bevel, chamfer,
Letting bugger, bastard,
Take the strain.

As for this thick air
Harried from post to pillar,
Those shadows inching
As cats unfurl their claws,
—

This worry of light
Struggling weakly,
The house takes it in:
A bustle of photons

Countless as death
In its courts of grass,
Millions of everything
Signifying nothing

But 'Keep your balance.
When poles dip
You need a blindfold
To cross clenched water,

To twist a handle
To turn the corner
From then to now
And what comes after.'

4 *Love it. Choose it*

Love it. Choose it. Whatever the words mean
Hauled from the moil, the tumult in the head
And heart. Walls will obey
The flock of paper thoughts you hang on them,
Curtains ring and skirr on runs of ice,
Glass let you in and out to watch
The green things growing with a steady violence
In what you call your garden, this room
Switch its cushions and its books about,
Light, insidious as a lover,
Lisp its morning catalogue of lies:
'This is for you. You make me what I am.
I was a dead thing till you quickened me.'

Now look: How rich I grow in your possessing,
Dreamy as a child with a button-box,
Showing you *you* in a cup splashed with flowers,
Dandyprats who prance a biscuit-tin,
A cat whose whiskers blacken at the root.
So much finery, Arabian stuff,
Caught in the prisms of your kitchen-sink
Where crumpled bubbles tear themselves to nothing
With the sound of a small fire in another room.
As you leave, the wicked light slips under
A carpet sewn with buttercups and clover,
And all is, as the rose in the desert, beautiful,
Though the door shuts, conversation becomes voices.

5 *No Dismissal*

The house, gathering its mass, says:
'Sursum corda.'

'But we have no hearts to lift. What are we?
Broken glass at night, and no glass broken,
Smells of warm fur at sleep's corner,
Crackle of bones in a grey floorboard.

In the high thrill of blood in a stranger's ear
We talk of what life once remembered
Which is not life. We are long ago,
And ride in dreams to the crop's firing.'

Beneath let and hindrance, in the leat
Where time flows: a scribble of dust-water,
Coccytus, conduit of tears, flux of pain:
Over the roof-tree, stars.

The house commands: 'Ite, missa est.'

'Where should we go? We are the faces
Annealed in glass at January's window,
A curtain slithering into costume,
The chance of blessing.

You are the root of our conspiracy,
House of Atreus, in which the hands of life
Wear unseen rings of marriage, and of mourning.
There is no dismissal for the wind.'

Painted hounds, leashed on a limed wall,
Each mouth unslaked, look to their quarry.
A crucifixion of withered nails,
The strained flesh dust,

The house groans:
'Asperge me, Domini, hyssopo,
Et mundabor.'

'There is no cleansing.'

6 Where is the Key?

Perhaps the house is full of awkward questions,
As if all questions weren't awkward:
A voice raised as the roof is raised,
Instinct with a question's anger.
Each stone interrogation
Cuts a blood-shape to the mark of Cain,
Each steady, single vowel lobs a grenade.
When the house was born, did it cry in the first wind
For its delivery from stone and clay,
Shaken, unhinged, gathering its baubles
To be the playthings of its own dark future?
When the stomach sickens, the gorge rises,
What is placated by this mess of sweetmeats?

Living here will be enough and more
To turn the hair white with grief.
Where is the key, the key to this parade
Of drawers and wardrobes. It is thinner now,
Though it always was a skeleton; unlocks
Life with a fluency you could despair of,
Knows the cosy terror of the habitual
The saved and sloughed-off skins
Which keep the curious shape of yesterday:
That walk to La Courante in driving snow,
The field of flax which opened a powdery eye.
When the key clicks, tugs, what is unlocked?
A coffin for the memory of memories.

7 *This Room, that Room*

This room is a hold of chattering teeth,
Split-pins shaken in a dry tumbler,
Light plugging your silhouette with flat knives
In the Cirque d'Hiver, the cold, the cold.
Washing hangs barbed on front-line wire
And air jolts to the high, explosive words
Tracked by the huge scream of their arrival.

That room is dense with a dead thing,
The surfaces are coffin-lids come off
And with them come the sticky skins of faces
Which might be where the flowers have had enough
And sent their candles drowning into wax.
The words? A Tennysonian murmuring:
Oh, oh, alas, alas, was, was.

This room believes in the belief of moths
And shows by the dark how to betrothe darkness
To the space where wing-chair and carpet sing
In their dull brilliance about gone sun.
I think the onset of Alzheimer's
Is making the words run in and out of their colours.
Quite charming, this prattle and confusion.

That room is locked black and sobbing.
The walls are dry bread and the floor is water
Where Alice wept when she was sent to bed.
When you touch wood, it turns to iron;
When you touch iron, the cold scalds you.
Listen. The words are kicking hard
At both sides of the door. Which is echo?

This room is where they put the other rooms.
Chekhov would have known the servants' names
Who pulled the rug from under the feet of time
And spread it out, a great big dirty dustsheet.
Light is loose-limbed over mottled marble
And words are tic-tac, miching-mallecho,
The chirping of dead mice from the underworld.

That room? Ah, that room is the last room,
The doorway swung on a brass fanfare of light,
The dustsheet flicked to a conjuror's hanky.
And here they are, so many he's and she's,
And all as pretty in their love and ugliness
As any thousand pictures. And the words
Are quite unnecessary.

8 *Numen Inest*

Ah, numen inest. At first, clouded,
They are always there, shapes of nothing:
A genie aching for the flanks of his bottle,
The stoppered stillness. From brush and deadwood,
A greenfield site with goalposts floating
In curious fog, a quick train's lazy rattle,
Come voices which insist on being missing,
Belonging out, but wanting in and blessing.

They need you, fly from you, the passers-by
At work in frost or green grazing:
A sharp retort in the copse, gruff talk
Lost in the hung stuff between earth and sky,
The burr and wobble of a plane pausing,
The trouble of branches going cark, crake.
Sound upon sound, adept for the long haul,
Easing its substance through the strongest wall

Until the crammed air's thick with frequencies:
Your mother preaching from her photograph,
Books at close argument, a singing cat
Jouncing its flap. All well-worn distances
Dissolve. The past unseals its epitaph
And that is this which certainly was that.
You feel the whole house brewing up a soul,
Choked on some filmy substance: a black hole

Where lives cook up a starry-gazy pie
And then are gone, dragging into its maw
Bruises and smiles, arpeggios and flowers.
It has a vampire's hunger, and an eye
That's unappeasable. It asks for *more*,
And gets it, as the tin parade of hours
Hastens its virgin tribute. What do days *do*?
They make a radioactive wraith of you.

9 *Let Things be Nice*

Let things be nice. Get out of bed the right side.
The floor is paved with sunflakes
Washing the floor and scraps of tissue clean.
It is that old-gold image, quattrocento.
Look into light: incursions into haloes
Make runes, mandalas, alphabets of time
Spent in a good place, with a good heart.
Today, the broken figurine can dance,
The clock purr, the chipped plate
Brim to its full circle. Under the willows
There is no danger on the unburnt bridge,
No parting of lovers, unstrung instruments.
It is time for a picnic on the far shores of your garden,
For loving Miss Bates, fostering the small endearments.

Now, add the colours of pastness to the proper names:
Madame Carrière, Saxifrage, Rosa Rubrifolia,
Obedient to that multiplicity of laws
By which invisible things flesh into substance.
Possession. Dispossession. What do you incubate
When a thorn drives deep, the heart grows callous?
Today, the windows of the house
Are all alive with the dark look of water.
The house is cupped for a draught of light,
The rinsing of veils and curtains.
What is reflected, but your own reflections
On the unspoken speech of brick and flowerhead?
Sloughed, dry as your skin;
Immortal as your memory.

10 Substance to Silence

Consort of viols, communion of the faithful?
Those handsome words; they work as the hands work,
Alert for lifting a little the dark's corners,
Holding nothing but air, but what does the air hold
Which lets go of nothing but is all union?
Single additions, which make the single full,
The knot of ribbons

Flown from fingers, field-talk, a swung necklace,
The chimney's cowl, the lacing flesh
Under its glittering coverlet of dust.
Sky, what do you purchase, when the pedlar
Unpacks the spectrum of old songs and snatches,
Juggling his feather-lights, his primaries,
Love, Death and Memory

Swung high, higher, caught at their vanishing
Point of departure, which is their arrival,
As, outside the gate the Vauxhall eases down
The box on its tonneau spilling a cornucopia
Of gifts once-given, twice-given, for ever given.
How pleased the dead are to see us, cheeks flushed,
Hands brown, and the names

Flying about like kisses, out into all,
The cat sprawled on its back in its gravelled fur,
A flower on the lawn of winter, each singleton
At the dear heart of all that's circular,
And all completed by what's fugitive
As the past opens its enormous future,
Reads us like a book,

And the oak-tree there, consonant and shapely
Where the stars can sing themselves invisibly
Into the white dark, lives by our whispering,
Cradling the air to one substantial mirage:
The deep, continuous thing on its endless way
From substance to silence. The House dreaming,
The many mansions,

Things living as they must, things that must live
By the light's traverse, holding their hands to it,
Whatever it is, which is light behind light,
And the book? The small print of a summer dress,
A tract of time. Hold out your hands
To the air; let it show you its finest detail,
It will bear your weight.

11 *Faces Come Thicker at Night*

Faces come thicker at night; a host of stars
Breaks uneven ground to a wild glister,
That different sameness which the cold flays bare
With each close look. Stars, which shine and shine,
Bothered by names and regimental numbers,
Are pesters of fine print against tired eyes.
They wear the look of something thought about
And make their own obedience to the gestural:
A purple codex lit by silver uncials.

In winter, not much need to name the grasses
Pushed through pea-shingle crunchy as packed snow.
What is, lives only in dim crevices,
Burrowed between its accident and substance.
Still, it's something to be startled by Orion,
Listen to a north wind, as Bevis and Mark did,
Pausing under an oak at the top of Home Field.
There are familiars: the dog-star blue with cold,
Pleiads and Hyades clouding into cataract.

We watch them into being, like the dead,
And, like the dead, they find the space too huge
For anything but whispers of forgiveness.
The house, black-backed, swells against open fields,
Hunted down by those long strides of starlight,
Those bright bows drawn at yet another venture.
Quick, throw it open, let its lanterns draw them,
Setting clean squares adrift for all that ghost-life,
This flurry of winter moths against glazed silence

For something in the bite and tang of time,
When moonlight hangs there with a medium's breath
To riddle, sift and screen them. Watching, waiting
Under these frosts, who can you turn from your door,
Such thousands hurry to your long house-warming?
There's Quince, the young white cat, whose corpse
Grinned in the ivy, who sleeps under his quince-tree,
And, dead, sits patient under this bright hood
For doors to open, all things to come home.

~

Scissors

So random, when people walk out of it like this,
Taking themselves and their flowers off to the cemetery,
Or, as now, lying down with their feet in the kitchen,
Head in the passage. It's a cold suspense.
And where to look? There's a hundred soft things,
Glittery things; they don't know where they are,

And can't be told any more, by the slightest of Angels,
How they came from Reigate, or that loft in Mayo.
Alas for the sepia people, their wits quite gone,
Labelled 'Aunt G at Filey' in such invisible ink.
And everything blue with cold in a crying November,
Especially now, when someone is asking for scissors,

Which after all, have to earn their severance pay,
Are silently, somewhere, whispering; working the dust,
Snipping the web into flyaway patterns of light.
The amputations, the deaths of a thousand cuts
For a daughter, a friend in Croydon, an Oxfam covenant,
A ticket for two for *Much Ado About Nothing*.

It's all so remorseless; they must have come into their own,
Setting about it with pure and bloodless precision.
Every drawer is a drawer for a kind of bridal,
And 'What will happen to me?' sighs the brass fender,
Who came naked into her world, and so must leave it
When the scissors have snapped each room to echo and
footstep.

You can hear them fuss, as blade slips over blade
With a kiss, with a kiss: a life folded over and over,
Grown paper-thin, and the cut sharp on the creases.
Are they moons, are they stars? Look, with a child's wonder
You can hold it up to the light, a white fretted window,
A seamless ribbon she steps through, smiles, and is gone.

A Twist of Water

It is always pouring itself away.
The scullery tap, when pipes knock and throb,
Wheels its head over a rod of water
With a sugar-stick twist, a cripple's twist,
Which only looks like a stronger, deeper light
Against all the other lights which have wandered
In from the garden.

And under it you can hold your rose of fingers
That have been places — a fistful without memory —
That curled in the womb, that have known
Such crush and pile of stuff, tasks under leaves,
Gestures in rooms where all the talk has run
Far out into yards and yards of blown fabric;
Known running water

Which always runs with a twist, though steady,
Its million beads fused in a shaking rod,
A racing pulse that your weeping hand flows through,
Feeling its lucid weight, the flap of a flame
Through which a finger can pass without much pain.
And still it is pouring itself away,
This trickster's wand

Waved over a tiny, white-lipped cut
Where the blood is rinsed in light, the blood
Which is not thicker than water, and quickly lost,
As if all those hurts which poison the wells
Could be cleansed by something brilliant as this,
Kind, held in a twist, like space, like time,
Which are pouring away.

When the Train Stops

When the trains stops, in a mid-afternoon,
And its motion continues, which is soon gone,
And the light stops too, your train of thought
Tripswitched, broken,

There was a house once, which you come from,
And one to visit, which is not quite the same
But will have its claim on you, as of right.
At your sudden waking,

Now, in the middle-distance, a house is there,
Drawn suddenly in stone on heated air.
The breakwater wall runs down to the gate,
There's a girl swinging.

And there it is: the completely known,
Flesh of your flesh, bone of bone,
Your washing blowing on a branching line,
All calling 'Mine',

And its mine is yours: the half-swung door,
Elder and Ash, the slight copse where
Your fingers pick for feather and skull,
The scuffed hill,

The curtains blowing across the dark —
You can read it easily as the book
By your side, which you will not read at all
Till you get there,

Which is not there. With sighs, relinquishings,
The undertow of some black trembling,
The carriage moves. There is nothing at all
To keep you . . .

OXFORD POETS

Fleur Adcock
Moniza Alvi
Kamau Brathwaite
Joseph Brodsky
Basil Bunting
Daniela Crăsnaru
W. H. Davies
Michael Donaghy
Keith Douglas
D. J. Enright
Roy Fisher
Ivor Gurney
David Harsent
Gwen Harwood
Anthony Hecht
Zbigniew Herbert
Thomas Kinsella
Brad Leithauser
Derek Mahon
Jamie McKendrick
Sean O'Brien
Peter Porter
Craig Raine
Zsuzsa Rakovszky
Henry Reed
Christopher Reid
Stephen Romer
Carole Satyamurti
Peter Scupham
Jo Shapcott
Penelope Shuttle
Anne Stevenson
George Szirtes
Grete Tartler
Edward Thomas
Charles Tomlinson
Marina Tsvetaeva
Chris Wallace-Crabbe
Hugo Williams